STRATOSPHERIC
OZONE
DEPLETION

STRATOSPHERIC OZONE DEPLETION

Ann M. Middlebrook
National Oceanic and Atmospheric Administration

Margaret A. Tolbert
University of Colorado
Department of Chemistry and Biochemistry and
Cooperative Institute for Research in Environmental Sciences

UNIVERSITY SCIENCE BOOKS
SAUSALITO, CALIFORNIA

University Science Books
univscibks@igc.org
www.uscibooks.com
Order Information: Telephone (703) 661-1572
                    Fax (703) 661-1501

Scientific director: Tom M.L. Wigley
Managing editor: Lucy Warner
Editor: Carol Rasmussen
Art and design: NCAR Image and Design Services
Cover design and composition: Craig Malone

The book's cover shows polar stratospheric clouds over McMurdo Station, Antarctica. The cover photograph is courtesy of David Hofmann, NOAA CMDL, Boulder, Colorado.

Library of Congress Cataloging-in-Publication Data

Middlebrook, Ann M., 1964–
Stratospheric ozone depletion / Ann M. Middlebrook, Margaret A. Tolbert.
     p. cm. – (Global change instruction program)
   Includes bibliographical references and index.
   Summary: Examines the phenomenon of ozone loss and considers its chemistry, causes, and prevention.
     ISBN 1-891389-10-6 (softcover : alk. paper)
     1. Ozone layer depletion. [1. Ozone layer. 2. Ozone layer depletion.]
I. Tolbert, Margaret A., 1957-II. Title. III. Series

QC879.712 .M53 2000
363.738′75–dc21

                                        99-058346

Printed in the United State of America
10 9 8 7 6 5 4 3 2 1

A Note on the Global Change Instruction Program

This series has been designed by college professors to fill an urgent need for interdisciplinary materials on global change. These materials are aimed at undergraduate students not majoring in science. The modular materials can be integrated into a number of existing courses—in earth sciences, biology, physics, astronomy, chemistry, meteorology, and the social sciences. They are written to capture the interest of the student who has little grounding in math and the technical aspects of science but whose intellectual curiosity is piqued by concern for the environment.

For a complete list of modules available in the Global Change Instruction Program, contact University Science Books, Sausalito, California, univscibks@igc.org. Information is also available on the World Wide Web at http://www.uscibooks.com/globdir.htm or http://www.ucar.edu/communications/gcip/.

Contents

Preface

As the 20th century draws to a close, the global impact of human activities, once inconceivable, is now a reality. These problems must be confronted and resolved to maintain the natural balance of the planet. In the past twenty years, the worldwide issue of ozone loss in the stratosphere has been identified, studied, and explained, and the international community has banded together to protect its mutual vital interests. During the next century, the results of this effort will be evident as the Antarctic ozone hole recedes. This module discusses the discovery of the Antarctic ozone hole, scientific studies of the processes that cause stratospheric ozone loss, and the international strategy to protect the ozone layer for future generations. Hopefully, the lessons learned from the ozone problem will pave the way for society to tackle other global and environmental issues. We dedicate this module to our children and their future.

A. M. M. and M. A. T.
Boulder, CO
January 2000

Acknowledgments

This instructional module has been produced by the Global
Change Instruction Program of the University Corporation
for Atmospheric Research, with support from the National
Science Foundation. Any opinions, findings, conclusions, or
recommendations expressed in this publication are those of
the authors and do not necessarily reflect the views of the
National Science Foundation.

This project was supported, in part, by the
National Science Foundation
Opinions expressed are those of the authors
and not necessarily those of the Foundation

STRATOSPHERIC
OZONE
DEPLETION

Introduction

In the last decade, there have been numerous reports of ozone loss in the upper atmosphere. Often the newspaper headlines and sound bites are too short to provide a complete picture and are simplified to tell only one side of the story. Some claim that ozone depletion is a monumental global problem, whereas others state that ozone loss is nonexistent. Even current advertising and science fiction movies are not immune to the dramatic story of ozone loss. A recent Maybelline advertisement promotes its Natural Defense™ make-up by stating, "You know what's happening to the ozone, imagine what it's doing to your skin." In the movie *RoboCop*, there is a futuristic advertisement for a sunblock with SPF 2000 because no ozone is left to stop the Sun's ultraviolet (UV) radiation from reaching Earth. Another sci-fi movie, *Highlander II*, depicts a physical barrier erected over Earth to provide UV protection because the ozone layer was completely destroyed.

Although the facts have sometimes been misrepresented, massive ozone loss over the Antarctic continent has been observed annually since 1979, significant ozone loss above the Arctic has been observed recently, and ozone has been eroding slowly above the United States and other populated regions. Why should we care about the ozone layer? Quite simply, ozone naturally shields the planet from incoming UV radiation from the Sun, and this radiation destroys deoxyribonucleic acid (DNA) and proteins in all living organisms. Increased UV radiation causes increased incidence of basal cell and squamous cell skin cancer and possibly melanoma, immune system deficiencies, and cataracts in humans. In addition, the productivity of ocean phytoplankton and certain crop plants is diminished by increased UV radiation. The effects of higher UV radiation are discussed further in another module, *Biological Consequences of Global Climate Change*.

An interesting and profound aspect of ozone depletion is that it is a global problem caused by human activities. Many people find it difficult to realize that humans can affect the environment on a global scale. In this module, we will describe the chemistry of ozone, how ozone destruction is attributed to human activity, and what society is doing to avert further damage to the ozone layer. At the end, we give a set of questions about ozone depletion.

I
The Stratosphere and the Ozone Layer

On the basis of temperature, the atmosphere is subdivided into different altitude regions like the layers of an onion (see Figure 1). Closest to the Earth's surface is the region called the troposphere. This region includes the air that we breathe and is where weather systems occur. The air pressure (and also the total number of molecules per unit volume) decreases almost exponentially with altitude so that roughly 90% of the mass of the atmosphere is in the troposphere and 50% of the mass is within the 5.5 km closest to the Earth.

In the troposphere, the air is generally well mixed vertically, so many gases released on Earth tend to be mixed throughout the troposphere. We can understand the vertical mixing of the troposphere by imagining what happens when a balloon of air at the ground is raised 1 km into the atmosphere. The balloon will expand because the pressure is lower at the higher altitude and will cool because it has to do work to expand. If no heat is allowed to be exchanged between the balloon and the surrounding air, we can calculate from thermodynamics what the new balloon temperature will be. In air that contains no water, the temperature will drop 10 K per km raised. If the balloon's new temperature is higher than that of the surrounding air at that altitude, the balloon will continue to rise. This example represents unstable air that is well mixed, a condition that is very common in the troposphere. If, however,

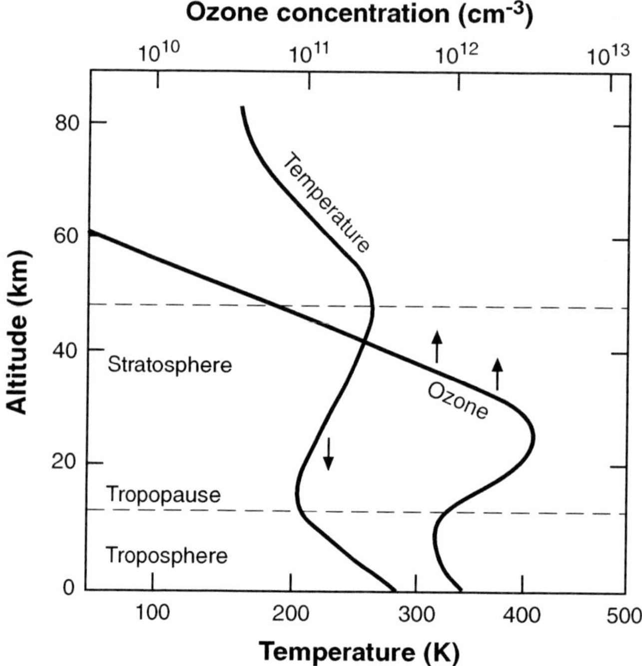

Figure 1. Variation of temperature (bottom axis) and ozone (top axis) with altitude from the surface of the Earth. In the troposphere, temperature decreases with altitude. In the stratosphere, temperature increases with altitude. The tropopause is located at the minimum temperature. Ozone concentrations maximize in the stratosphere. The ozone concentration is shown on a log scale in units of the number of molecules per cubic cm (#/cm^3). Adapted from Watson, R.T., M.A. Geller, R.S. Stolarski, and R.F. Hampson: Present State of Knowledge of the Upper Atmosphere: An Assessment Report. *NASA Reference Publication 1162, NASA, Washington, D.C., 1986, p. 20.*

the balloon's new temperature is lower than that of the surrounding air, the balloon will descend to its original position. This represents very stable air with very little vertical mixing. This situation, where cold air is below warm air, is often referred to as a temperature inversion and is a contributing factor to smog.

As seen in Figure 1, above the troposphere the temperature begins to increase as the altitude increases. This is due to the absorption of light energy by ozone. The altitude where the lowest temperatures occur in the lower atmosphere defines the top of the troposphere and is called the tropopause; above that is the region called the stratosphere. Because the stratosphere is a giant inversion layer with warm air above cold air, there is very slow vertical mixing in this region. In very broad terms, the global atmospheric circulation consists of upwelling into the stratosphere near the equator and downward transport of stratospheric air near the poles. The tropopause height is 16–18 km near the equator and 8–12 km near the poles.

A diagram of ozone concentration as a function of altitude is included in Figure 1. Ozone

Good Ozone and Bad Ozone

In Figure 1, there is a slight increase in ozone concentrations at altitudes very close to ground level. High concentrations of ozone are often found in the troposphere in polluted areas such as Los Angeles, California. Although ozone near the ground will also absorb DNA-damaging ultraviolet radiation, it is toxic to plants and humans (see *Biological Consequences of Global Climate Change*). Therefore, ozone can be thought of as "good" in the stratosphere and "bad" in the troposphere (where we would breathe it). Consequently, efforts should be made to reduce the amount of tropospheric ozone formed by pollution and maintain the ozone layer in the stratosphere.

Chemicals Named in This Module

Br	bromine
BrO	bromine monoxide
$BrONO_2$	bromine nitrate
BrO_x	bromine oxides
CCl_2FCClF_2	1,1,2-trichloro-1,2,2-trifluoroethane (CFC-113)
CCl_4	carbon tetrachloride (CFC-10)
$CFCl_3$	trichlorofluoromethane (CFC-11)
CF_2Cl_2	dichloro-difluoromethane (CFC-12)
CH_2BrCl	bromochloromethane
$CHClF_2$	chlorodifluoromethane (HCFC-22)
CH_3Br	methyl bromide
CH_3Cl	methyl chloride
CH_3CCl_3	methyl chloroform
CH_4	methane
Cl	chlorine atom
ClO	chlorine monoxide
$ClOO$	unstable chlorine dioxide
$(ClO)_2$	chlorine monoxide dimer
$ClONO_2$	chlorine nitrate
ClO_x	chlorine oxides
Cl_2	chlorine molecule
CO_2	carbon dioxide
HBr	hydrogen bromide
HCl	hydrochloric acid
HNO_3	nitric acid
HO_2	hydroperoxyl radical
HO_x	hydrogen oxides
H_2O	water
H_2SO_4	sulfuric acid
M	any molecule
NO	nitric oxide
NO_2	nitrogen dioxide
NO_3	nitrate radical
NO_x	nitrogen oxides
N_2O	nitrous oxide
N_2O_5	dinitrogen pentoxide
$NaCl$	sodium chloride
O	oxygen atom
$O*$	highly energized oxygen atom
OCS	carbonyl sulfide
OH	hydroxyl radical
O_2	oxygen molecule
O_3	ozone
SO_2	sulfur dioxide

concentrations reach a maximum in the stratosphere, forming the so-called ozone layer. Although ozone is a vital component of the stratosphere, the actual amount of ozone is very small. If all of the ozone in the atmosphere were compressed to the pressure at sea level, the ozone would form a layer only 3 mm thick, compared with 8,500 m if the entire atmosphere were similarly compressed.

The chemical composition of the stratosphere is quite different from that of the troposphere. For example, water concentrations are much lower in the stratosphere because water condenses into clouds in the troposphere, and these rain out. In addition, water rising through the tropical tropopause tends to freeze out due to the extremely cold temperatures there. Only a small amount of water vapor gets through this "cold trap." Another source of stratospheric water vapor is from the oxidation of CH_4 gas (see list of chemical names and formulas on p. 3) released on Earth. Because so little water vapor is present in the stratosphere, very few clouds form there. However, these clouds play a crucial role in the ozone story.

Many of the molecules released on Earth do not reach the stratosphere because they are soluble (can dissolve) in water and return to the surface in precipitation or are broken down by chemical reactions in the troposphere. These include such molecules as HCl and NaCl. Both HCl, from erupting volcanoes, and NaCl, released into the atmosphere by ocean wave-breaking, are very soluble in water, so they are not a significant source of chlorine to the stratosphere. Some molecules, such as hydrocarbons emitted from plants and from incomplete combustion of fossil fuels, react in the troposphere with OH before they can reach the stratosphere.

The only molecules that do reach the stratosphere are those that are insoluble in water and also are chemically inert (or unreactive). A partial list of these compounds includes nitrous oxide (N_2O), carbonyl sulfide (OCS), methane (CH_4), chlorofluorocarbons (CFCs), and halons (carbon-based molecules containing bromine). Some of these species are natural and some are anthropogenic (generated by human activity). Although these molecules do not react in the troposphere, when they reach the stratosphere they absorb UV light and break apart. Such reactions in which molecules absorb light and fall apart are called photolysis (or photochemical) reactions. As we shall show below, photolysis of these molecules in the stratosphere affects ozone concentrations.

The intensity of solar radiation also varies between the troposphere and stratosphere. Both gases and particles absorb and scatter radiation, preventing the full amount of sunlight from

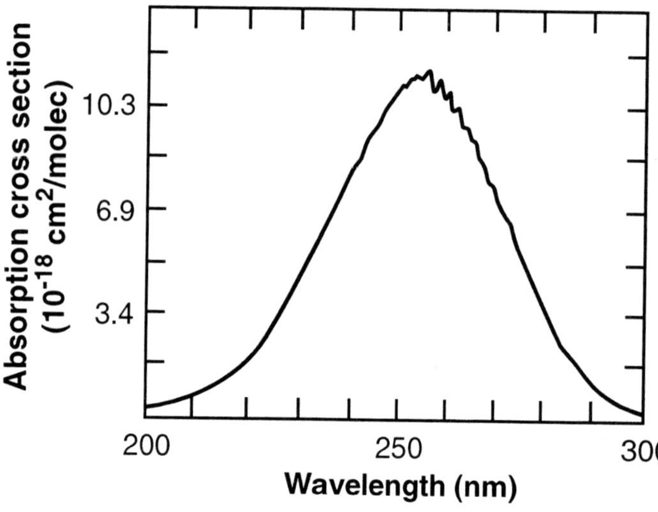

Figure 2. The ozone absorption spectrum showing where ozone filters UV radiation. The large absorption band of ozone between 200 and 300 nm is called the Hartley band, named after its discoverer, John Hartley. From Griggs, M.: Absorption coefficients of ozone in the ultraviolet and visible regions. Journal of Chemical Physics 49, 1968, pp. 857–859. © 1968, American Institute of Physics. Reprinted by permission.

Table 1: Ultraviolet and Visible Radiation

Name	Approx. Wavelengths	Comments
UV-C	200–280 nm	Extremely damaging to DNA, but absorbed by both molecular oxygen and ozone
UV-B	280–320 nm	Very damaging to DNA, sensitive to ozone changes
UV-A	320–400 nm	Suntan region (some DNA damage)
Visible	400–700 nm	Detectable with human eyes

reaching the ground. The amount of light at a given wavelength reaching the planet's surface depends on the gases and particles present in the atmosphere. A complete discussion of the Earth's radiation budget is described in the module *The Sun-Earth System*. Here, it is important to note that ozone absorbs UV radiation (wavelengths, λ, between 200 and 300 nm, see Figure 2), prevent-ing high intensities at these wavelengths from reaching the ground. Table 1 shows the approximate wavelength regions of ultraviolet and visible light. If less ozone is present in the stratosphere, less UV light is absorbed there, so more reaches the ground. In fact, an increase in UV radiation has been observed as a function of decreasing ozone concentrations (see Figure 3).

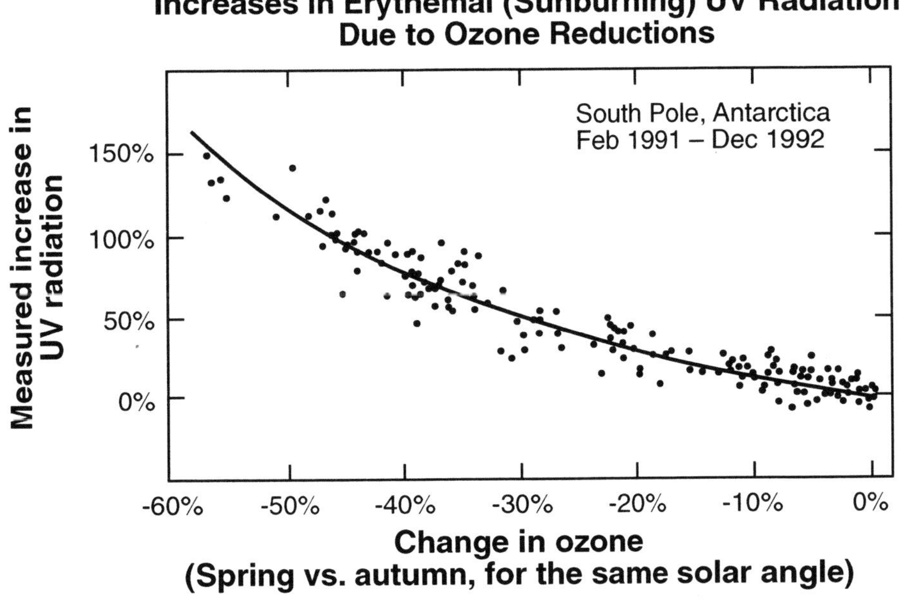

Figure 3. Measured increase in ultraviolet (UV) radiation at the South Pole as a function of change in ozone. From WMO Global Ozone Research and Monitoring Project: Scientific Assessment of Ozone Depletion: 1994. WMO, Geneva, Switzerland, No. 37, 1994, p. xxxiii.

The Dobson spectrophotometer for measuring ozone is a ground-based instrument invented by the English scientist George M.B. Dobson in 1927. It measures the column amount of ozone (total number of ozone molecules above a square centimeter of the Earth's surface) by measuring the amount of ultraviolet light absorbed by the atmosphere (see Figure 4). The longest Dobson record is from the station in Arosa, Switzerland, where ozone has been measured continuously since 1932. In 1957, the International Geophysical Year, a network of 85 Dobson stations was established to measure global ozone. This network has provided long-term data showing significant ozone loss on a global scale over the past 25 years.

The unit for measuring the column amount of ozone is called the Dobson Unit (DU), where one DU equals 2.7×10^{16} molecules per cm^2. This number was chosen because it represents 100 times the height in mm of ozone per cm^2 over Earth if all of the ozone were brought to the same atmospheric pressure as at sea level. So, 3 mm of ozone per cm^2 equals 300 DU. Other units used to measure actual concentration are number of molecules per cm^3 as shown in Figure 1; the mixing ratio of ozone to air in ppmv, ppbv, or pptv (1 part per million, billion, or trillion by volume is 1 molecule per 10^6, 10^9, or 10^{12} total molecules); and partial pressure in millibars (1 mb = 9.87×10^{-4} atmospheres) or milliPascals (1 mPa = 9.87×10^{-9} atmospheres), which is also the total pressure times the mixing ratio.

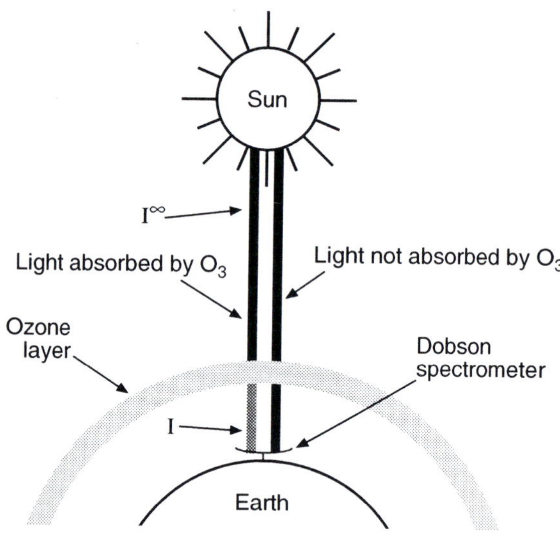

Figure 4. Schematic diagram showing how ozone is measured by Dobson spectrometers. Because ozone transmits the visible wavelengths and absorbs the ultraviolet, the amount of ozone can be determined by comparing differences in absorption at two wavelengths. Beer's Law is used to calculate the amount of light transmitted (T) through an absorbing species, like ozone. A simplified example of this is shown, where the intensity of the Sun without any absorption by ozone is I^{∞} and the intensity reaching the ground through the ozone layer is I. The fraction of available light reaching the ground is given by

$$T = \frac{I}{I^{\infty}} = \exp\left[-N \times \sigma(\lambda)\right]$$

where I^{∞} = the solar intensity above the ozone layer, I = the solar intensity reaching the ground, N = the column amount of ozone molecules (molec/cm^2), and $\sigma(\lambda)$ = the absorption cross-section (cm^2/molec) at the wavelength = λ (nm).

II
Ozone Chemistry

Ozone, a molecule consisting of three oxygen atoms, was first discovered in the 1830s by the German scientist Christian Schönbein. He identified a new compound in laboratory experiments using oxygen, and named the molecule "ozein," meaning "to smell" in Greek. In 1881, John Hartley experimented with ozone and found that it strongly absorbed ultraviolet light (see Figure 2). He compared the absorption spectrum of ozone to the spectrum of sunlight as seen from the Earth's surface and found that they matched exactly.

But where in the atmosphere was the ozone that absorbed the Sun's ultraviolet light? Ozone measurements in the troposphere in the early 1900s showed that not enough ozone was there to explain the observed column amount. Thus, the ozone must be higher up in the atmosphere. The vertical profile of ozone was finally established by measurements of Gotz and Dobson in the 1930s, who showed that the ozone was located mainly in a layer about 22 km above the Earth. The fact that ozone concentrations peaked in the stratosphere suggested a photochemical (sunlight-driven) source for ozone. In 1930, Sidney Chapman (an English scientist) proposed a four-reaction mechanism to explain the global distribution of stratospheric ozone. The ozone is produced by the photolysis of oxygen molecules into oxygen atoms, shown in Reaction (1), followed by the reaction of one oxygen atom with an oxygen molecule:

$$O_2 + UV\ light \rightarrow O + O \qquad \lambda < 242\ nm \qquad (1)$$
$$O + O_2 + M \rightarrow O_3 + M \qquad (2)$$

Net: $2\ O_2 + UV\ light \rightarrow O + O_3$

In Reaction (1), the ultraviolet light required to dissociate the O_2 molecules must have a wavelength less than 242 nm. The oxygen atoms formed in Reaction (1) react with an oxygen molecule to form ozone in Reaction (2). The third molecule, M, is needed to remove the excess energy and can be any other molecule in the atmosphere. Note that the concentration of M (or number of M molecules per unit volume) decreases as the pressure decreases with increasing altitude. When Reactions (1) and (2) are added together, they form a single net reaction, shown below the solid line. The net reaction indicates that oxygen atoms and ozone are produced from oxygen molecules and light. Ozone can be destroyed by photolysis [Reaction (3)] and by reaction with an oxygen atom:

$$O_3 + light \rightarrow O_2 + O \qquad (3)$$
$$O + O_3 \rightarrow 2\ O_2 \qquad (4)$$

Net: $2\ O_3 + light \rightarrow 3\ O_2$

We see that Reaction (3) produces an oxygen atom, which can recombine with an oxygen molecule to reform ozone as in Reaction (2). Therefore, Reaction (3) does not necessarily result in ozone destruction unless it is followed by Reaction (4). The net reaction for Reactions (3) and (4) shows that ozone is destroyed by light to regenerate oxygen molecules.

In the above Chapman mechanism, it can be seen that sunlight is an important requirement for the formation of ozone. It turns out, however, that the maximum amount of ozone is not found in the tropics where the solar radiation is the highest, but rather at higher latitudes in the spring. Measurements of the global column amount of ozone over the Earth as a function of month of the year (prior to the ozone hole formation) are shown in Figure 5. The contour lines indicate the ozone concentrations at that latitude and time of year. It can be seen that there is a peak in ozone amount at 75° N latitude in March (460 DU) and a smaller peak at 60°S latitude in October (400 DU). The peaks in ozone at high latitudes occur because the ozone concentrations are a strong function of both production chemistry and meteorological transport. Recall that air is transported poleward and downward from the tropics, bringing ozone to the poles where it is long-lived due to low levels of sunlight. This transport is strongest in the winter, causing ozone to accumulate at high latitudes and thus maximize in the spring.

Other factors also can influence global ozone levels. Because Reactions (1) and (3) involve sunlight, variations in the amount of solar radiation influence global ozone. The UV light from the Sun varies in an 11-year cycle, with ozone levels decreasing 1–2% from the maximum to the minimum. However, ozone losses discussed later in this module far exceed 1–2%, and their timing does not correlate with the solar cycle.

The Chapman mechanism for ozone does remarkably well in predicting where in the atmosphere ozone can be found, with a peak in the lower stratosphere. However, the amount of ozone predicted by the Chapman mechanism is greater than the amount of stratospheric ozone actually observed. This is because, while Reactions (1) and (2) are the only net source of oxygen atoms and ozone, there are many ways in which ozone can be destroyed besides Reaction (4). For example, a Dutch scientist, Paul Crutzen, recognized in the early 1970s that ozone is catalytically destroyed in a cycle involving nitrogen oxides or NO_X (NO_X = NO + NO_2). For his pioneering work on ozone chemistry, Crutzen was a co-recipient of the 1995

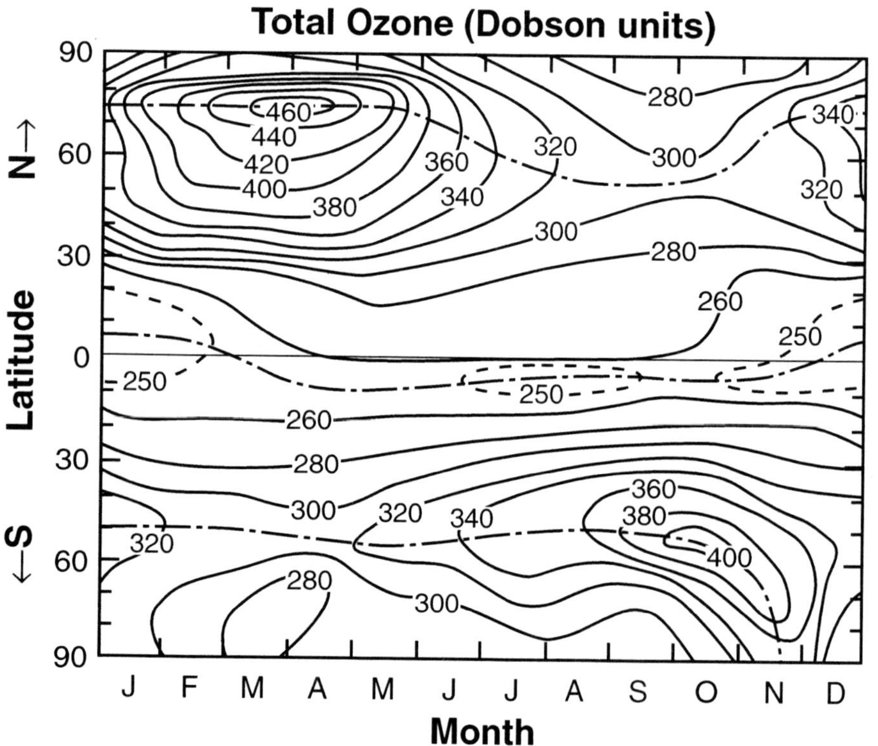

Figure 5. The global distribution of the column amount of ozone before the formation of the ozone hole. The column amounts of O_3 were derived from ground-based measurements. From London, J.: Radiative energy sources and sinks in the stratosphere and mesosphere, pp. 703–721, in: Nicolet, M., and A.C. Aikin (eds), Proceedings of the NATO Advanced Study Institute on Atmospheric Ozone: Its Variations and Human Influences, U.S. Dept. of Transportation, Washington, D.C., 1980.

Nobel Prize in chemistry. The catalytic cycle that he proposed is shown below:

$$NO + O_3 \rightarrow NO_2 + O_2 \quad (5)$$
$$O + NO_2 \rightarrow NO + O_2 \quad (6)$$

$$\text{Net: } O + O_3 \rightarrow 2\,O_2$$

The net reaction above shows that an oxygen atom and an ozone molecule form two oxygen molecules. Interestingly, the net reaction does not use up any NO_x molecules because when Reactions (5) and (6) are added, neither NO nor NO_2 is produced or destroyed. Thus, this ozone loss cycle is catalyzed by NO_x because the NO consumed in Reaction (5) is regenerated in Reaction (6) and the NO_2 produced by Reaction (5) is consumed in Reaction (6). Therefore, this cycle can repeat many times as long as some NO_x is available for reaction. NO_x occurs both naturally and, to a lesser extent, anthropogenically in the stratosphere. A natural source is the oxidation of N_2O released in the troposphere by bacteria in soil and oceanic microorganisms. An anthropogenic source of N_2O is fertilization. N_2O molecules do not photolyze or react in the troposphere, and thus are able to reach the stratosphere. Once there, some N_2O reacts with a highly energized oxygen atom, O^*, to form NO. Nitrogen oxides may also be injected directly into the stratosphere by exhaust from high-flying aircraft (see below), as first recognized by the U.S. scientist Harold Johnston at the same time that Crutzen proposed the above NO_x ozone loss cycle.

Also in the 1970s, Richard Stolarski and Ralph Cicerone (both scientists from the United States) proposed that chlorine from solid fuel rockets could cause ozone depletion via the catalytic cycle:

$$Cl + O_3 \rightarrow ClO + O_2 \quad (7)$$
$$O + ClO \rightarrow Cl + O_2 \quad (8)$$

$$\text{Net: } O + O_3 \rightarrow 2\,O_2$$

This scheme is also catalytic, with one atom of chlorine being able to destroy 100,000 molecules of ozone. Eventually, the chlorine atoms react with CH_4 to form HCl, temporarily ending this ozone destruction cycle. Fortunately, the amount of chlorine emitted into the stratosphere from rockets and space shuttle exhaust is very small. However, chlorine became recognized for potentially destroying ozone.

The main source of chlorine oxides ($ClO_x = Cl + ClO$) to the stratosphere was recognized in 1974 by the U.S. scientists Mario Molina and Sherwood Rowland to be CFCs. For their important discovery of the link between CFCs and ozone depletion, Molina and Rowland were co-recipients of the 1995 Nobel Prize in chemistry. CFCs are molecules containing only carbon, chlorine, and fluorine, and are known to be chemically inert. In fact, it was their inert nature that made them ideally suited for use as aerosol propellants, refrigerants, foam-blowing agents, and cleaning solvents. Because they are so inert, they do not react in the troposphere. The rapid mixing of the troposphere helps these molecules to spread uniformly there. Recall that most air entering the stratosphere rises through the tropical tropopause. Therefore, CFCs rise in the tropics and are spread to both the northern and southern polar stratospheres even though most sources are in the Northern Hemisphere. We know that CFCs reach the stratosphere because they have been observed there for many years (see Figure 6). Scientists also have measured the concentrations of various chlorine species in the stratosphere and have determined that over 80% of the chlorine compounds are anthropogenic (see Figure 7). In addition to the above chlorine reactions with ozone, other trace species, such as bromine oxides ($BrO_x = Br + BrO$) and hydrogen oxides ($HO_x = OH + HO_2$), can also lead to catalytic ozone destruction. The relative importance of these species depends on altitude and latitude.

Figure 6. Measurement of CFC-11 (CFCl₃) from the ground up to the stratosphere at midlatitudes in the Northern Hemisphere. These data were obtained from five separate studies over the dates indicated. Note that the concentration decreases once CFCl₃ is above the ozone layer. The ozone layer is blocking the UV light needed to photo-chemically break apart the CFCl₃ via:

$$CFCl_3 + UV\ light \rightarrow CFCl_2 + Cl$$

Once the gas drifts above the ozone layer, it releases chlorine atoms. From WMO Global Ozone Research and Monitoring Project: Atmospheric Ozone: Assessment of Our Understanding of the Processes Controlling Its Present Distribution and Change. *No. 16, WMO, Geneva, Switzerland, 1985, p. 634.*

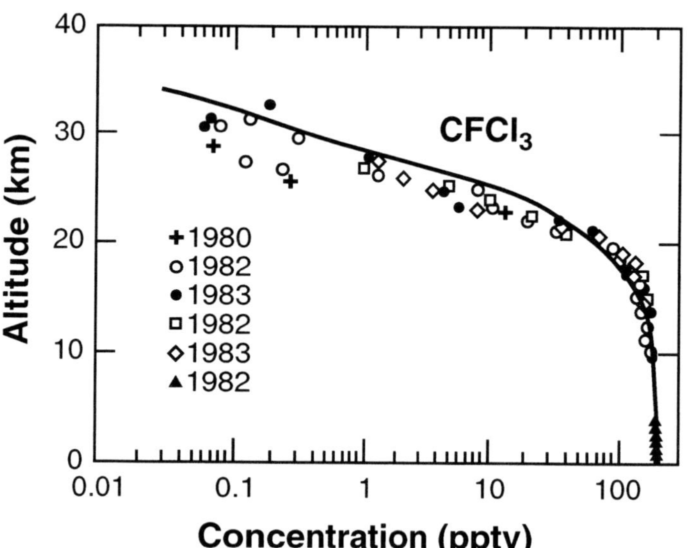

Figure 7. Pie chart of chlorine-containing compounds in the stratosphere, showing that 82% of the chlorine sources are anthropogenic. From WMO Global Ozone Research and Monitoring Project: Scientific Assessment of Ozone Depletion: 1994. *No. 37, WMO, Geneva, Switzerland, 1994, p. xxix.*

Primary Sources of Chlorine Entering the Stratosphere

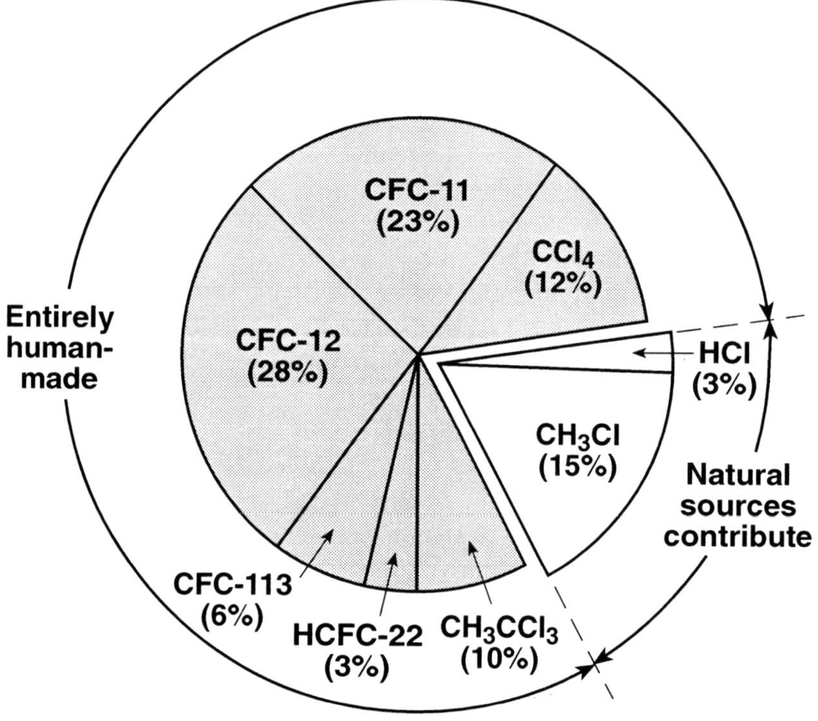

How Ozone Allowed Life to Evolve on the Surface of the Earth

It is believed that early in the planet's history, life existed only in the oceans where the harmful UV radiation was absorbed and scattered by wave action, bubbles, and debris floating in the water (Figure 8). Photosynthesis in ocean plants then converted CO_2 and H_2O into O_2 and the sugars that provide the energy to make the plants grow. When the ocean became saturated with O_2, the amount of oxygen in the atmosphere gradually increased. Gas-phase oxygen was consumed by two mechanisms: (1) new life forms used O_2 and sugars to produce CO_2, H_2O, and energy in the reverse-photosynthesis process called respiration, and (2) O_3 was formed by the photochemical destruction of O_2. As concentrations of ozone in the atmosphere began to increase, less UV radiation reached the Earth's surface. Because the ozone absorbed the DNA-damaging radiation, life could begin to exist on the Earth's surface rather than just under water.

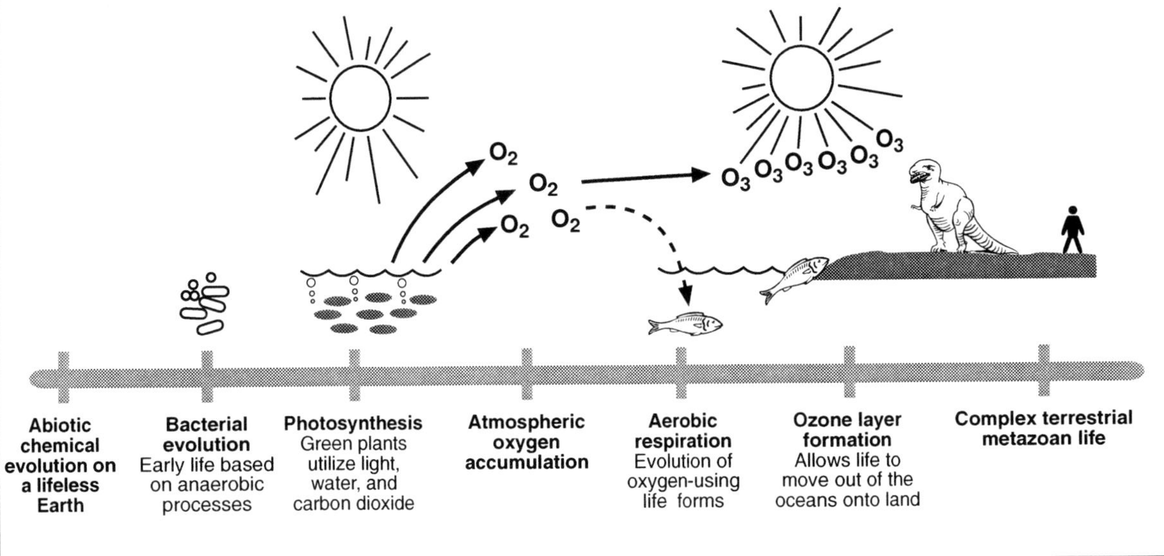

Figure 8. How life evolved on Earth. The oxygen produced by underwater plants eventually accumulated in the atmosphere. When UV light broke apart the oxygen, ozone was formed. After the ozone layer was formed, UV light was prevented from reaching the ground. Then, plants and animals could exist on the surface of the planet. From R. P. Turco, Earth Under Siege: Air Pollution and Global Change, *Oxford Press, p. 86, 1993.*

III

The Ozone Hole

When gas-phase catalytic ozone destruction cycles involving NO_x, ClO_x, BrO_x, and HO_x were first included in photochemical atmospheric models, global ozone levels were predicted to drop 5–10% over the next 100 years. This decrease was predicted because CFCs were increasing rapidly in concentration, causing the amount of chlorine in the atmosphere to increase.

The total column amount of ozone had been measured by the British Antarctic Survey (BAS) since 1957 at Halley Bay (76°S latitude), using a Dobson spectrophotometer. Beginning in 1979, Joe Farman and coworkers at BAS began to observe substantial decreases in ozone in the early spring. At the same time, the Total Ozone Mapping Spectrometer (TOMS) and the Solar Backscatter Ultraviolet (SBUV) instruments aboard NASA's Nimbus 7 satellite were also making ozone measurements. However, these instruments were not reporting the large ozone decreases that the ground-based researchers measured. So the BAS scientists waited years to make certain that their observations were correct before publishing their results. Their astounding report that up to 30% of the column ozone was depleted in the span of a few weeks caused the satellite scientists at NASA to reanalyze their data. They soon discovered that their computer program, which had been designed to disregard abnormally low ozone data (thought to be influenced by cosmic rays), was actually removing most of the data for that time period. After the computer algorithm was corrected, the satellite instruments confirmed the ozone measurements made by the BAS team. Dobson measurements at U.S. and Japanese Antarctic stations since the 1960s also show the dramatic ozone

loss after the late 1970s (see Figure 9).

Subsequent measurements of ozone concentration as a function of altitude showed exactly where the ozone was being destroyed. Figure 10 shows a typical ozone "hole." The ozone profile during the dark polar winter (23 August 1993 data) resembles the profile shown in Figure 1, with ozone concentrations peaking at roughly 17 km. However, by only seven weeks later (12 October 1993), almost all the ozone between 14 and 19 km was destroyed. The integrated amount of ozone in this profile is 91 DU, the lowest ever measured. For comparison, the lowest column amount in 1992 was 105 DU, and the

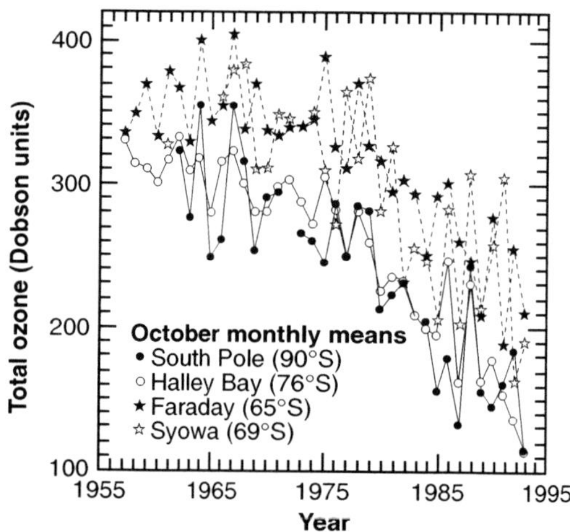

Figure 9. Average ozone concentrations in the month of October over the South Pole, Halley Bay, Faraday, and Syowa, Antarctica. All stations show that ozone levels have been decreasing since roughly 1975. From WMO Global Ozone Research and Monitoring Project: Scientific Assessment of Ozone Depletion: 1994. No. 37, WMO, Geneva, Switzerland, 1994, p. 1.44. By Shanklin, J., T. Ito, and D. Hofmann.

October column amount was about 300 DU until the late 1970s.

The ozone concentration record in the satellite data showed that the region where ozone loss occurred over Antarctica was a rough circle or "hole" centered near the South Pole. The ozone "hole" was found to be deeper and to last longer every other year until 1989, 1990, and 1991, when the holes were comparable. On Oct. 8, 1993, the lowest column ozone ever measured by TOMS (85 DU) was recorded above McMurdo Station (78°S). In 1998, the area covered by the ozone hole was the largest on record. When scientists first tried to explain these losses, they soon realized that they could not do so by the known gas-phase chemical reactions, such as Reactions (5) through (8). So, a search began for alternative (and better) explanations, and for data to test alternative hypotheses.

Explaining the Ozone Hole

Since the models containing known gas-phase chemistry failed to predict the dramatic loss of ozone over Antarctica, it was clear that something was missing. Three possible explanations emerged as those studying the ozone layer came to grips with the news of the hole. One theory was related to variations in solar energy. As mentioned above, the Sun has an 11-year cycle of increasing and then decreasing activity. When the solar activity is high, increased amounts of NO_x are produced. The NO_x could reduce the ozone concentrations by Reactions (5) and (6). This theory was soon disproved when scientists measured low NO_x concentrations during the ozone hole rather than the elevated ones required by the solar theory. Also, the losses observed over Antarctica are much greater than the 1–2% expected from variations in the solar cycle. In addition, the solar theory predicted ozone loss via Reactions (5) and (6) to occur at 40 km altitude, whereas the actual ozone loss occurred between 15 and 20 km (see Figure 10).

Another theory to explain the ozone hole involved changes in atmospheric dynamics (the

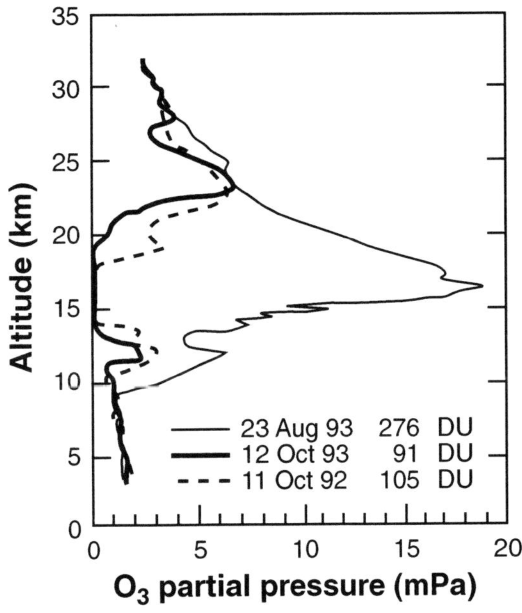

Figure 10. Ozone profile measured in 1992 and 1993 at the South Pole Station, Antarctica. On August 23, the ozone concentration profile (right-hand line) appears normal. On October 12, ozone concentrations between 14 and 19 km were below the instrument's detection limit, illustrating vertical location of the ozone hole. For comparison, the ozone profile on October 11, 1992, is also shown. From Hofmann, D.J., S.J. Oltmans, J.A. Lathrop, J.M. Harris, and H. Voemel: Record low ozone at the South Pole in the spring of 1993. Geophysical Research Letters 21, 1994, pp. 421–424.

air's circulation patterns). At the beginning of the austral winter, a large, circular vortex develops over the Antarctic continent (see Figure 11). This vortex effectively isolates the air above Antarctica from midlatitude air, preventing ozone-rich air from the tropics from mixing with the polar air. In addition to this isolation, the dynamical theory for the ozone hole suggested that the special cold conditions of the Antarctic caused upwelling when the Sun returned, moving ozone-poor air from lower to higher altitudes. Later, however, field observations showed that the N_2O concentrations were lower inside the vortex than outside, indicating that the vortex air is older and had been transported downward from above. Thus, air moves downward inside the polar vortex, not upward as needed in the dynamic (circulation) theory.

The third theory to explain the ozone hole was based on chemistry. Catalytic chain reactions caused by unusually high levels of ClO_x were proposed to be the cause of the enhanced ozone loss. Ground-based field measurements led by U.S. scientist Susan Solomon in the spring of 1986 in Antarctica showed greatly enhanced levels of ClO above that expected from known chemistry. This was confirmed by in-flight measurements of ClO in the stratosphere above Antarctica in 1987, where low ozone levels were found in air with high ClO levels. The data from flights into the vortex (see Figure 12) clearly show that the ozone concentrations decrease wherever ClO increases and vice versa. This relationship of low ozone with high ClO is referred to as the "smoking gun" of ozone destruction because it definitively

demonstrated that chlorine was the culprit.

Although it was now clear that anthropogenic chlorine was indeed to blame for the ozone loss, the exact mechanism was still not known. In the winter and early spring above Antarctica, strong sunlight is unavailable to create oxygen atoms, which are needed for ozone loss via the chlorine catalytic cycle in Reactions (7) and (8). Thus, another mechanism was needed to recycle the ClO back to Cl atoms. Mario and Luisa Molina proposed that the ClO reacts with itself to form a dimer, $(ClO)_2$, which in turn photolyzes into Cl atoms for ozone destruction:

$$2 \times [Cl + O_3 \rightarrow ClO + O_2] \qquad (7)$$
$$ClO + ClO + M \rightarrow (ClO)_2 + M \qquad (9)$$
$$(ClO)_2 + light \rightarrow Cl + ClOO \qquad (10)$$
$$ClOO + M \rightarrow Cl + O_2 + M \qquad (11)$$

Net: $2\,O_3 + light \rightarrow 3\,O_2$

Because Reaction (9) needs two ClO molecules, this ozone-destroying cycle is only effective if the concentration of ClO is very high. This mechanism is now thought to account for roughly 80% of the ozone destroyed above Antarctica.

Under typical midlatitude conditions, ClO rapidly reacts with NO_2 to form $ClONO_2$:

$$ClO + NO_2 + M \rightarrow ClONO_2 + M \qquad (12)$$

When this happens, the chlorine is attached to a nonreactive, or reservoir, molecule ($ClONO_2$) that does not deplete ozone. Therefore, for ClO to deplete ozone, it is helpful if NO_2 levels are low. These special conditions of high ClO and low NO_2 are now recognized to occur in the winter polar regions due to the presence of polar stratospheric clouds.

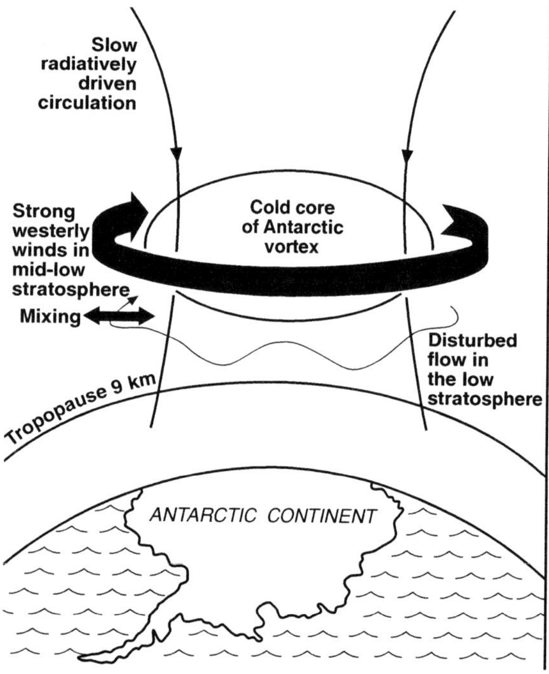

Figure 11. Schematic diagram of the Antarctic vortex formed during the winter, showing the isolation of polar air from the rest of the stratosphere. From Wayne, R.P.: Chemistry of Atmospheres, 2nd ed., p. 184. © 1991, Oxford University Press. Reprinted by permission.

The Role of Polar Stratospheric Clouds

Polar stratospheric clouds (PSCs) have been observed in polar winters above Antarctica for decades. Because little water vapor reaches the stratosphere, the temperature must be very cold

(<187 K) for water-ice particles to condense and clouds to form. However, each year above Antarctica and also frequently above the Arctic Ocean, stratospheric temperatures do become that low in the winter. Another type of PSC that contains crystals of nitric acid (HNO_3) and ice may also form in the stratosphere. These nitric acid/ice particles form at slightly higher temperatures (<195 K) than are needed for the pure ice clouds. Because they do not require such a very low temperature for their formation, these clouds are more common than ice clouds, and are found every year over both poles during the winter.

It was first suggested by U.S. scientist Susan Solomon and coworkers that heterogeneous chemistry (requiring a gas and a surface) occurring on PSC particles could be the missing piece of the puzzle to explain the ozone hole. It is now recognized that PSC particles catalyze the reaction:

$$ClONO_2 + HCl \rightarrow Cl_2 + HNO_3 \tag{13}$$

For her key scientific insights into explaining the cause of the ozone hole and advancing the understanding of the global ozone layer, Solomon was awarded the 1999 National Medal of Science by President Clinton.

Reaction (13), occurring on the surfaces of PSC particles, converts the stable, ozone-friendly molecules $ClONO_2$ and HCl into a form more threatening to O_3. Photolysis of the Cl_2 molecules at polar sunrise produces chlorine atoms:

$$Cl_2 + light \rightarrow 2\ Cl \tag{14}$$

These chlorine atoms are then capable of catalytic ozone destruction via the Molina and Molina chlorine cycle, Reactions (7) and (9) through (11). Therefore, the net effect of Reactions (13) and (14) is to activate chlorine so that it can destroy ozone. Laboratory experiments showed that Reaction (13) was slow in the gas phase. However, after the discovery of the ozone hole, several research groups quickly designed experiments to investigate Reaction (13) on surfaces representative of PSC particles and found that the presence of an ice surface made the reaction go much faster.

In addition to activating chlorine, Reaction (13) also forms HNO_3, which remains incorporated in the PSC particles. This conversion removes NO_x from the polar winter strato-

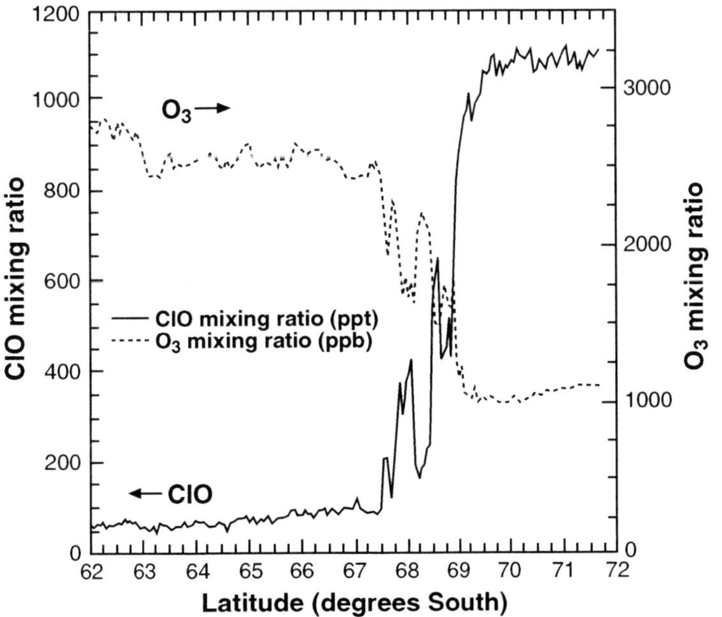

Figure 12. Ozone/ClO "smoking gun" relationship measured above Antarctica in September 1987. As the ER-2 aircraft making the measurements approached the ozone hole region, ozone concentrations decreased and chlorine monoxide concentrations increased. From Anderson, J.G., W.H. Brune, and M.H. Proffitt: Ozone destruction by chlorine radicals within the Antarctic Vortex: The spatial and temporal evolution of ClO-O_3 anticorrelation based on in situ ER-2 data. Journal of Geophysical Research *94, 1989, pp. 11465–11479.*

sphere. If the cloud particles grow large enough to fall, the HNO_3 falls with them. Sedimentation of HNO_3 associated with PSC particles can cause permanent removal of nitrogen species from the stratosphere, the so-called denitrification. Low levels of NO_X in the Antarctic stratosphere prevent the sequestering of active chlorine (ClO) back into inactive forms ($ClONO_2$) via Reaction (12). This overall sequence of events is shown schematically in Figure 13. Therefore, the net effect of the reaction of $ClONO_2$ with HCl (Reaction 13) on the surfaces of PSC particles is to produce unusually high levels of ClO_X and unusually low levels of NO_X.

Although chlorine was found to be vital for polar ozone depletion, laboratory and field measurements now indicate that bromine species also play an important role in destroying ozone. Another way of converting ClO into Cl atoms involves BrO_X:

$$Cl + O_3 \rightarrow ClO + O_2 \qquad (7)$$
$$Br + O_3 \rightarrow BrO + O_2 \qquad (15)$$
$$BrO + ClO \rightarrow Br + ClOO \qquad (16)$$
$$ClOO + M \rightarrow Cl + O_2 + M \qquad (11)$$

$$\text{Net: } 2\,O_3 \rightarrow 3\,O_2$$

In fact, on an atom-for-atom basis, bromine is 50–100 times more destructive of ozone than is chlorine. This is partly because bromine compounds usually exist in the active BrO_X forms rather than the non-ozone-destroying forms ($BrONO_2$ and HBr), and thus are immediately ready for ozone destruction. The main source of bromine in the stratosphere is methyl bromide, emitted naturally by oceanic biological activity and from anthropogenic activities such as soil fumigation and biomass burning. Other potentially important anthropogenic bromine sources are halons, used in fire extinguishers. Currently, bromine gases are less than 1% as abundant as chlorine gases. Even so, ozone destruction by the bromine/chlorine cycle, Reactions (7), (15), (16), and (11), is thought to account for about 20% of the ozone lost in the polar regions.

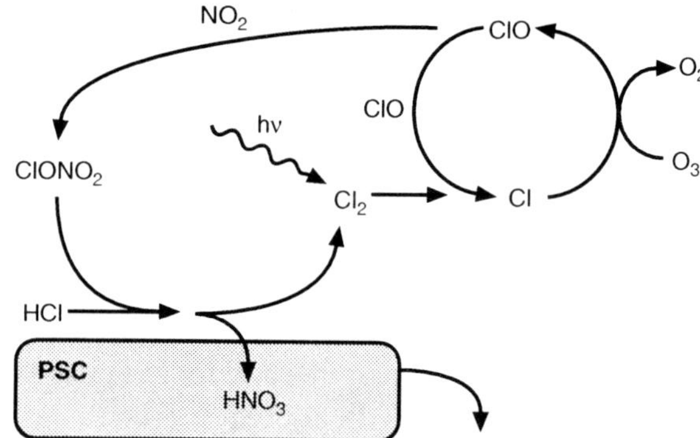

Figure 13. Schematic representation of PSC processing. Reservoir chlorine molecules, $ClONO_2$, react with HCl, also a chlorine reservoir, on PSC surfaces to form HNO_3 and Cl_2. Sunlight (hv) causes the Cl_2 to photolyze into Cl atoms, which catalytically destroy ozone. The catalytic cycle continues until ClO reacts with NO_2 to reform $ClONO_2$. NO_2 is produced by the photolysis of gas-phase HNO_3. However, HNO_3 during the polar night may be permanently removed if the PSC particle containing HNO_3 falls out of the stratosphere. This results in low levels of NO_2, and thus higher levels of ClO for O_3 destruction.

Summary of Conditions for the Ozone Hole

The three requirements for massive ozone loss in the stratosphere are now recognized to be (1) chlorine and bromine in the atmosphere, (2) cold temperatures and particle surfaces for heterogeneous chemistry, and (3) modest amounts of sunlight. This is shown schematically in Figure 14. These three conditions are all met in the unique environment of the Antarctic stratosphere. The polar vortex that forms each winter above Antarctica is important for isolating the stratospheric air, which allows the air to become very cold during the winter and reduces the influx of ozone-rich air from the tropics. The low temperatures allow PSCs to form in the winter/spring and thus support the heterogeneous chemistry shown in Reaction (13). Sunlight is needed to drive the actual ozone-loss cycles such as Reactions (10) and (14). The PSCs also lead to denitrification, which prevents the active ClO from reforming inactive $ClONO_2$. At the altitude of PSCs, the ozone is completely destroyed. Only when the polar vortex breaks down in the late spring/early summer does ozone from lower latitudes fill in the hole. It is important to note that of the three requirements for ozone loss, the only one that has significantly changed since the 1950s is the amount of chlorine and bromine in the atmosphere. Thus, it is the increasing levels of these halogens that have caused the regular appearance of the Antartic ozone hole.

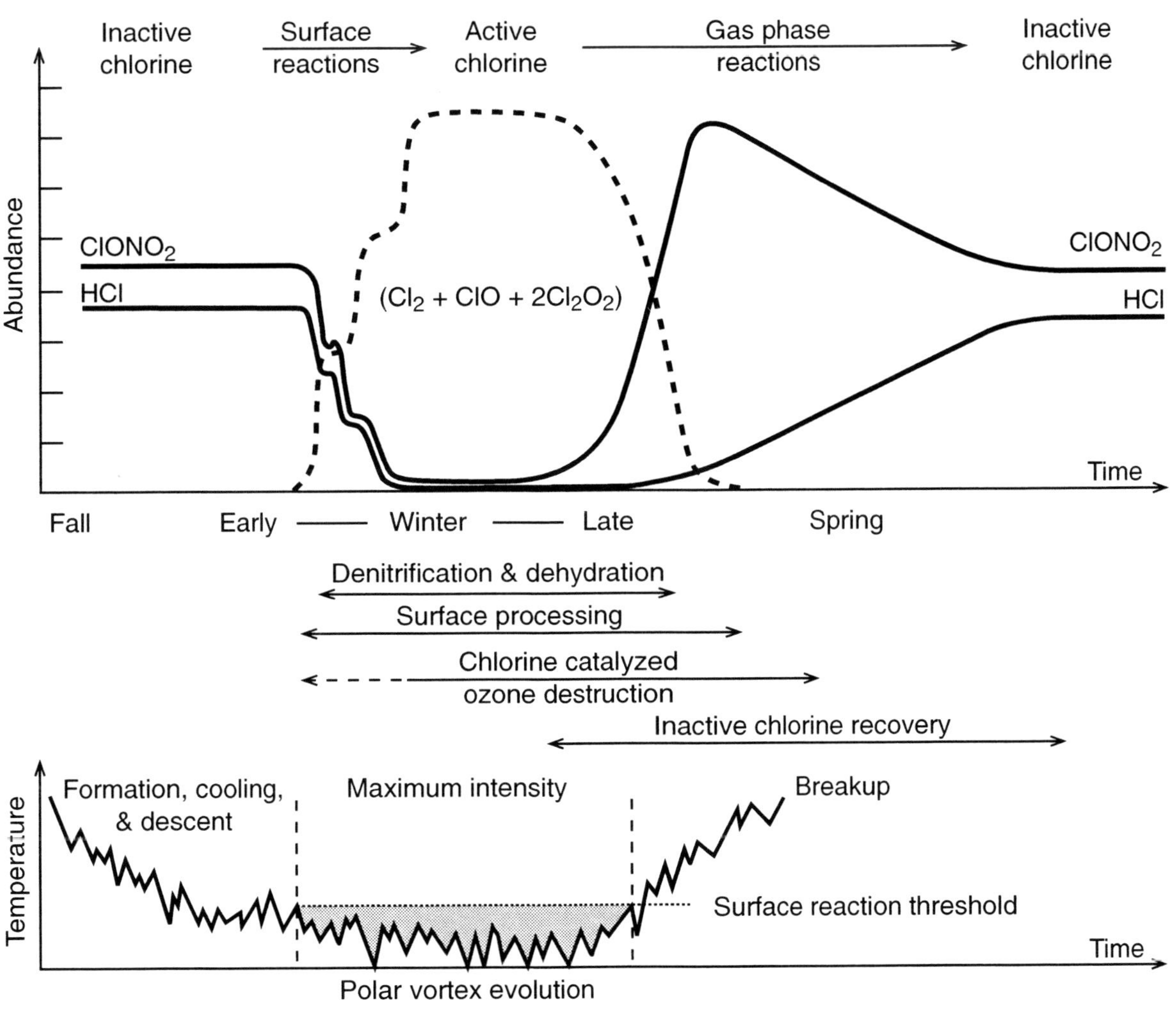

Figure 14. Schematic diagram of the chemistry that causes the ozone hole. During the cold, dark polar winter, PSCs form, and the reservoir chlorine compounds ClONO₂ and HCl react on their surfaces. With very little sunlight, the Cl₂ is pho-tolyzed into Cl atoms that destroy ozone. As the temperature increases in the late spring, the PSCs evaporate and the reser-voir reactions slow down, resulting in increased ClONO₂ and HCl concentrations and replenishment of ozone. From WMO Global Ozone Research and Monitoring Project: Scientific Assessment of Ozone Depletion: 1994. *No. 37, WMO, Gene-va, Switzerland, 1994, p. 3.4.*

IV

Arctic and Global Ozone Destruction

Arctic Ozone Destruction

Why doesn't an ozone hole appear over the Arctic? The same amount of total chlorine and bromine exists in the northern polar regions as above Antarctica, and both regions have sunlight during the spring. An important difference between the two areas is temperature, which affects how many PSCs form, how long they persist, and how much they overlap with the returning sunlight. Data for 1980–88 show that minimum polar stratospheric temperatures are always low enough for ice PSCs in the south polar region, but only occasionally low enough for them in northern regions (see Figure 15). In the north polar region there are more nitric acid/ice PSCs than pure ice PSCs. The main dif-

ference between the two types of PSCs is how well they remove nitric acid from the gas phase. Although the nitric acid/ice particles contain more nitric acid by weight than do the pure ice particles, the ice particles are more effective at permanently removing nitric acid from the stratosphere. This is because the ice particles are larger, so they fall faster out of the stratosphere, taking the nitric acid with them. If the temperatures above the Arctic were lower for longer periods of time, more ice PSC particles could form, thus removing more nitric acid and leading to more ozone loss. Even in the absence of ice PSCs, chlorine activation via Reaction (13) still occurs efficiently on the background stratospheric sulfate aerosols at low temperatures. Thus, low temperatures in the Arctic would favor ozone loss with-

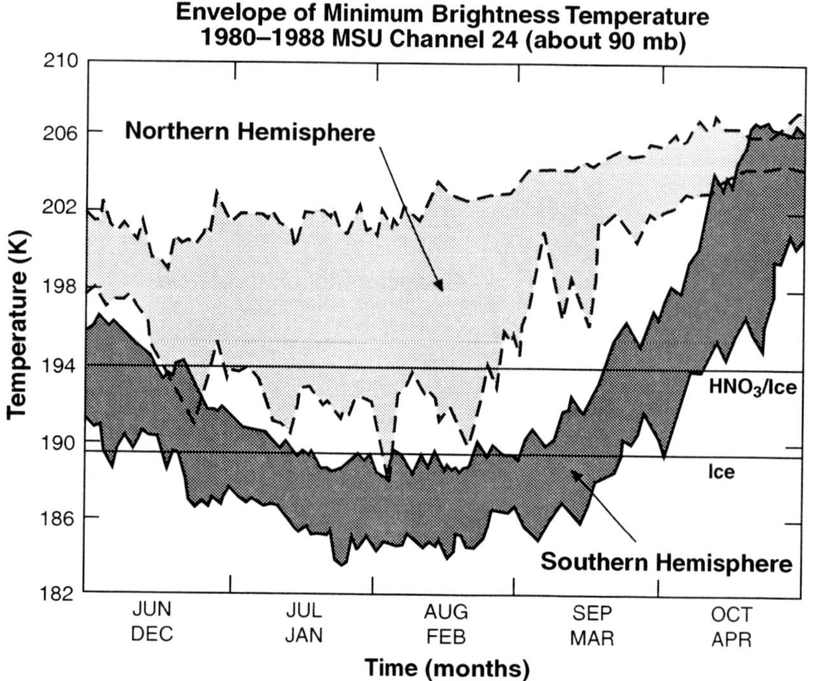

Figure 15. Diagram showing the envelope of minimum temperatures in both the northern and southern polar regions of the stratosphere, during each hemisphere's winter and spring. The approximate temperature thresholds for HNO₃/ice and ice are depicted as horizontal lines. Note that the Southern Hemisphere is much colder than the Northern Hemisphere and thus supports the formation of more PSCs. From WMO Global Ozone Research and Monitoring Project: Scientific Assessment of Stratospheric Ozone: 1989. No. 20, WMO, Geneva, Switzerland, *1989, p. 83.*

out permanent removal of nitric acid.

The low temperatures and low ozone levels over Antarctica are sustained by its very strong vortex circulation in winter. Although the Arctic stratosphere also develops a vortex circulation pattern during the winter, it is not as strong as the one formed over Antarctica, and air from the warmer, lower latitudes often moves into the Arctic region. This inhibits the Arctic air from becoming as cold as in the Antarctic. However, if ozone loss were to occur, the weaker Arctic vortex could then transport that ozone-poor air to lower latitudes over populated areas.

During the winter of 1991/92, scientists found that all of the chlorine above the Arctic was essentially ready for catalytic ozone destruction. Fortunately, massive ozone loss did not occur at that time because the temperatures did not remain low throughout the winter and spring. However, we can reasonably expect ozone loss in the northern polar regions to occur during colder winters. In fact, during the winter of 1994/95, temperatures in the Arctic were the lowest in 30 years, and scientists observed extremely low ozone concentrations—column amounts as much as 50% below normal.

The temperatures in the Arctic winter 1999/2000 were also extremely low, with the region of temperature low enough to form PSCs being the largest recorded in over 40 years of stratospheric analyses. Ozone loss of over 50% occurred during this Arctic winter at altitudes near 18 km. While dramatic, the Arctic ozone loss is still not as severe as that over Antarctica, and thus these recent events have been called the Arctic "half-a-hole." The appearance of the Arctic "half-a-hole" proves that ozone loss can and probably will occur over northern latitudes as long as the polar winters remain very cold and chlorine is present in the stratosphere.

Global Ozone Destruction

The polar ozone holes are reasonably simple to explain through heterogeneous chemistry and atmospheric dynamics. Field measurements have shown that the chlorine causing the dramatic ozone loss comes from anthropogenic sources. Hence, we are responsible for the ozone holes. Therefore, we have introduced policies to control the release of chlorine into the atmosphere (see Section V). However, it is now apparent that stratospheric ozone loss is not confined to the cold polar regions, but is also occurring globally. Data from the TOMS instrument on the Nimbus 7 satellite from 1978 to 1990 show an overall decrease of up to 0.8% per year in ozone at midlatitudes (see Figure 16). In addition, the latest data from TOMS show that the midlatitude ozone concentrations reached their lowest levels on record in 1992 and 1993 after the eruption of Mt. Pinatubo. We do not fully understand the mechanisms causing this global ozone loss. However, recent work suggests that heterogeneous reactions, like those

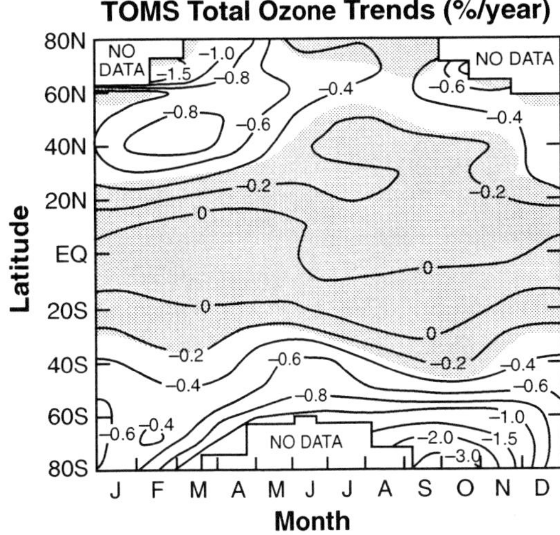

Figure 16. Global trends in total ozone amounts from 1978 to 1990. Percent change per year was deduced from TOMS satellite data as a function of latitude and time of the year. The decrease in polar ozone during the austral spring (September– November) and decreases in ozone at northern mid-latitudes (up to 0.8% per year) are observed. From Stolarski, R.S., P. Bloomfield, R.D. McPeters, and J.R. Herman: Total ozone trends deduced from Nimbus 7 TOMS data. Geophysical Research Letters 18, 1991, pp. 1015–1018. © American Geophysical Union.

occurring on PSC particles, could take place on stratospheric sulfate aerosols (SSAs), which become more widespread after explosive eruptions like Mt. Pinatubo. Therefore, heterogeneous reactions could also be contributing to the observed global-scale decrease in ozone levels.

SSAs are composed of concentrated solutions of sulfuric acid (H_2SO_4) in water. The background levels of these aerosols (i.e., in the absence of volcanic eruptions) are thought to arise naturally from the oxidation of OCS released at ground level by microorganisms. However, the amount of sulfuric acid in the stratosphere can increase by up to two orders of magnitude following a major volcanic eruption. For example, the eruption of El Chichón in 1982 injected SO_2 into the stratosphere, which rapidly oxidized to H_2SO_4 and became globally distributed. Ozone loss was detected at midlatitudes after the El Chichón eruption, possibly due to heterogeneous chemistry on SSAs. In 1991, the eruption of Mt. Pinatubo injected 2–3 times as much SO_2 into the stratosphere as did El Chichón. The combined increase in aerosol surface area due to the Mt. Pinatubo eruption and increase in chlorine in the stratosphere may be responsible for the record low ozone concentrations observed recently by TOMS.

The role of sulfuric acid particles is twofold. First, they can promote chlorine activation at low temperatures via Reaction (13), which is known to be important on PSC particles. Second, the SSAs promote the lowering of NO_x through Reactions (17) and (18):

$$NO_2 + NO_3 \leftrightarrow N_2O_5 \text{ (gas phase)} \quad (17)$$
$$N_2O_5 + H_2O \rightarrow 2\,HNO_3 \text{ (on SSAs)} \quad (18)$$

Lower levels of NO_x can indirectly lead to increases in ClO by slowing the formation of $ClONO_2$ via Reaction (12). Recent inclusions of Reactions (13), (17), and (18) into stratospheric models show that heterogeneous chemistry on SSAs could cause enhanced ozone loss due to elevated ClO and decreased NO_x.

It is important to note that the ozone loss following volcanic eruptions is not due to chlo-

rine injected from the volcanoes. Field measurements clearly show that HCl is not enhanced after such eruptions because it is washed out before reaching the stratosphere. Rather, the volcanic eruptions cause an increase in the surface area of SSAs, which then directly or indirectly activate chlorine for ozone loss. The main source of chlorine to the stratosphere is the anthropogenic CFCs. Thus, volcanoes alone do not cause ozone loss, but volcanic SSAs in conjunction with anthropogenic chlorine and bromine do. Therefore, part of the long-term downward trend in global ozone is due to the higher levels of chlorine in combination with heterogeneous SSA chemistry.

Although increased chlorine levels and heterogeneous reactions on SSAs explain part of the trend in ozone, they cannot account for all of the observed ozone loss. Some of the loss can be accounted for when temperature and concentration fluctuations are included in the models. Several theories have been put forward to help explain the remaining discrepancy between modeled and measured ozone. One is the dilution of air when the ozone hole dissipates in the late spring, i.e., mixing of low-ozone air from higher to lower latitudes. Another idea is that air from the midlatitudes is processed through the polar regions, causing ozone to be destroyed in an assembly-line fashion. The problem with these two theories is that neither is likely to produce the observed magnitude of the ozone loss, or to explain the roughly equal ozone loss observed in the two hemispheres. Another possible contribution to global ozone loss is a change in the dynamics of the atmosphere so that less ozone is transported to midlatitudes.

Although none of these theories alone can fully explain the current downward trend of global ozone loss, a combination may be responsible. In the future, atmospheric scientists will be trying to explain these ozone trends and predict further changes to the ozone layer in light of perturbations both natural (e.g., volcanic) and anthropogenic (e.g., chlorine and bromine compounds that replace CFCs and high-flying aircraft—see Section V).

V

Ozone Depletion and Policy

Since the 1950s, CFCs have been used in refrigerators, in aerosol spray cans, as cleaning solvents, and as foam-blowing agents. After Molina and Rowland showed in the 1970s that chlorine from CFCs could destroy ozone, CFC use in spray cans (aerosols) for hairspray, deodorant, and paint was restricted in the United States, Canada, Norway, and Sweden. Figure 17 shows how this ban affected worldwide CFC production and consumption. Overall, the production of CFCs has steadily increased from the 1960s to the mid-1980s except for the ten-year period directly after CFCs in aerosols were banned. The consumption pie charts show that the relative usage of CFCs in aerosols decreased between 1974 and 1988. However, usage for other applications continued to rise after aerosol usage dwindled, causing the overall production rate to climb during the 1980s. This is an example of poor regulations as the *use* of CFCs in aerosols was restricted instead of the CFCs.

The growing use of CFCs and the hypothesis that CFCs cause ozone depletion prompted many nations of the world to sign the Vienna Convention for the Protection of the Ozone Layer in 1985. Although the Vienna Convention

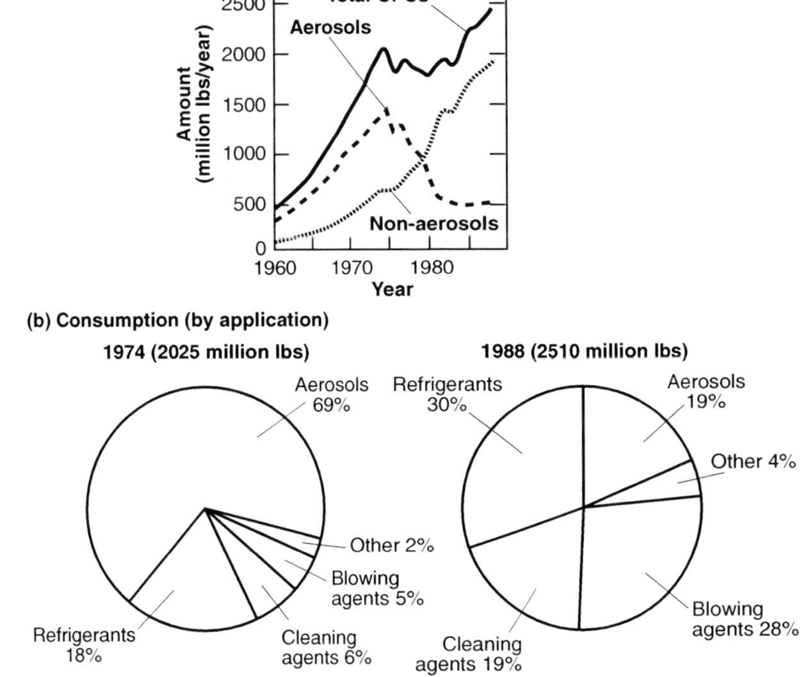

Figure 17. CFC use plots showing total world production from 1960 to 1988 and pie charts of the applications of CFCs in 1974 and in 1988. The amount of CFCs used in aerosol spray cans has decreased since the late 1970s. However, CFCs are increasingly being used as refrigerants, cleaning agents, and blowing agents. Although the relative amount of aerosol use has decreased, the total production of CFCs was still increasing in the 1980s due to the other applications. From McFarland, M.: Chlorofluorocarbons and ozone. Environmental Science and Technology 23, 1989, pp. 1203–1205. © 1989, American Chemical Society.

did not include explicit restrictions on CFCs, it called for future regulatory actions and scientific understanding of the ozone layer, CFCs, and halons. More importantly, it recognized that ozone depletion was an international issue and that the policy questions it raised would be resolved by the international community. Thus, it set the stage and provided the framework for the Montreal Protocol of 1987 (see below). Furthermore, it was prepared before the discovery of the ozone hole and, in fact, before any ozone loss was observed in the "real" atmosphere.

Following the framework of the Vienna Convention, the discovery of the ozone hole led to the formation of the Montreal Protocol on Substances That Deplete the Ozone Layer, which officially limited the production and use of CFCs. The countries that signed the original document in 1987 agreed to freeze CFC production and use at the 1986 rates by the year 1989, and to cut CFC production and use by 50% over the next ten years. After the protocol was adopted, it was determined that the rates proposed for CFC reduction were not rapid enough to substantially decrease the chlorine loading of the atmosphere. In fact, under the original protocol that was established in Montreal in 1987, chlorine levels would still be increasing by the end of the 21st century (see Figure 18).

After scientific evidence demonstrated that the ozone hole was caused by the chlorine from CFCs and that midlatitude ozone loss was also occurring, the 1990 London amendments were added to the Montreal Protocol to accelerate the reduction of the original CFCs listed to 100% reduction, i.e., a complete phaseout, by the year 2000. The London amendments also provided a timetable for 100% phaseout of other ozone-destroying compounds, including a more complete list of CFCs, halons, carbon tetrachloride, and methyl chloroform. After the London amendments were adopted, the chlorine and bromine atmospheric levels were expected to peak by 2020 and not return to pre–ozone hole levels until the end of the 21st century, as shown in Figure 18.

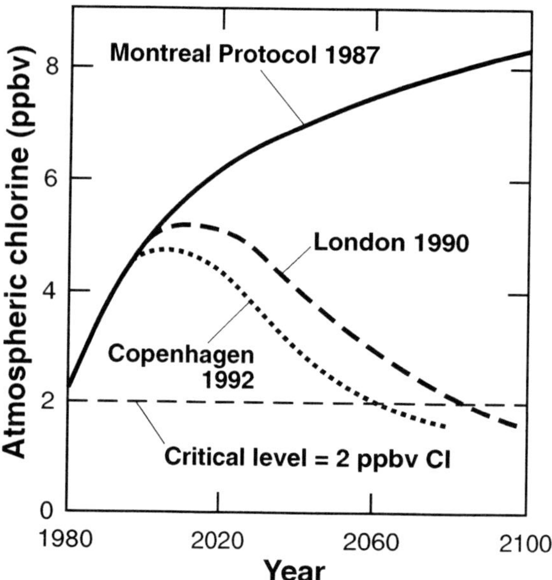

Figure 18. Atmospheric chlorine loading vs. year for different scenarios. The critical level of 2 ppbv is the amount of chlorine when the ozone hole began to form. Not shown are the predictions for chlorine levels without any restrictions. From Turco, R.P.: Earth Under Siege: Air Pollution and Global Change, *p. 434. © 1996, Oxford University Press, Inc. Reprinted by permission.*

Research continued, and the scientific understanding of ozone depletion progressed. Detailed model calculations showed that if the CFCs were phased out four years earlier, the time for chlorine concentrations to return to pre–ozone hole levels would be reduced by 20 years. (Compare the London and Copenhagen calculations in Figure 18.) Furthermore, in the winter of 1991/92, scientists who were monitoring ozone loss above the Arctic reported to the public the presence of unusually high amounts of ClO over the northern high- to midlatitudes. Fortunately, the winter did not remain cold enough for the active chlorine to replenish itself via heterogeneous reactions.

The predictions that chlorine concentrations could be more rapidly returned to pre–ozone hole levels coupled with the immediate possibility of ozone loss over populated regions prompted the addition of the Copenhagen amendments in 1992. These amendments provided for the complete phaseout of CFCs, carbon tetrachloride, and methyl chloroform by the year 1996. Furthermore, hydrochlorofluorocarbons (HCFCs), which were being developed to temporarily replace CFCs, were added to the list for phaseout by the year 2030 because they also contained chlorine that could destroy ozone. Since HCFCs react in the troposphere with OH, most HCFCs do not reach the stratosphere intact like the CFCs. Therefore, the threat to the ozone layer is not as serious from HCFCs as it is from CFCs. Finally, these amendments recognized the potential for the bromine-containing compound methyl bromide, used widely as a fumigant, to destroy ozone. As shown in Reactions (7), (15), (16), and (11), bromine can destroy ozone in a synergistic cycle with chlorine. However, decreases in chlorine levels and increases in bromine levels could increase the importance of bromine-catalyzed ozone loss. The Copenhagen amendments provided a limit of the methyl bromide production rate to 1991 levels beginning in 1995.

Other amendments to the Montreal Protocol were drafted at the 1997 Montreal meeting to commemorate the tenth anniversary of the original protocol. There, the parties agreed to accelerate the complete phaseout of methyl bromide production by five years to 2005 for developed countries and to set 2015 as the date of phaseout for developing countries. They also agreed to set up a licensing system for import and export of substances controlled by the protocol.

When the parties met again in Cairo in 1998, one of the main topics of discussion was how to make the Montreal Protocol policies consistent with the 1997 Kyoto Protocol on global warming. Some of the ozone replacement compounds, such as hydrofluorocarbons (HFCs) and perfluorocarbons (completely fluorinated hydrocarbons), absorb infrared radiation and could create a conflict between protecting the ozone layer and preventing the buildup of greenhouse gases. Furthermore, global climate models indicate that a warmer troposphere causes cooler temperatures in the stratosphere, which could

affect the ozone layer. Although it was agreed that policies for the two protocols needed to be coordinated, no new amendments were generated from this meeting.

In 1999, the parties met in Beijing and drafted amendments that mainly focused on HCFCs. They agreed to freeze production of these compounds to 1989 levels in 2004 for developed countries and to 2015 levels in 2016 for developing countries. Complete phaseout of HCFCs will occur in 2020 for developed countries and in 2040 for developing countries. Furthermore, trading of these compounds will be banned unless the trading countries ratify the 1992 Copenhagen amendments. The Beijing amendments also called for complete phaseout of a new compound, bromochloromethane (CH_2BrCl), for all countries by 2002.

For the amendments to enter into force, at least 20 countries must ratify, accept, or approve them. As of March 2000, all but the Beijing amendments have been ratified by the minimum number of countries. Text of the Protocol and amendments and the current ratification status are available to view on the United Nations Environment Programme website at www.unep.ch/ozone.

So, where would we be without CFC regulations? Without any restrictions, the chlorine levels would continue to increase by about 1 ppbv per decade (see Figure 18). If more countries were to become industrialized in the absence of any CFC regulations, the rate of increase would become even larger. By the year 2050, we would have had at least 10.5 ppbv more of chlorine in the atmosphere, which is off the scale of Figure 18. At the time the ozone hole appeared, chlorine concentrations in the atmosphere were about 2 ppbv. With the 1992 Copenhagen amendments, the chlorine is expected to peak at 4.3 ppbv near the year 2000 and to decrease to pre–ozone hole concentrations by 2060. Therefore, under the most stringent, current CFC regulations, we can reasonably expect the ozone hole formed by chlorine cycles to be a problem for decades. However, with the CFC regulations in place, the chlorine

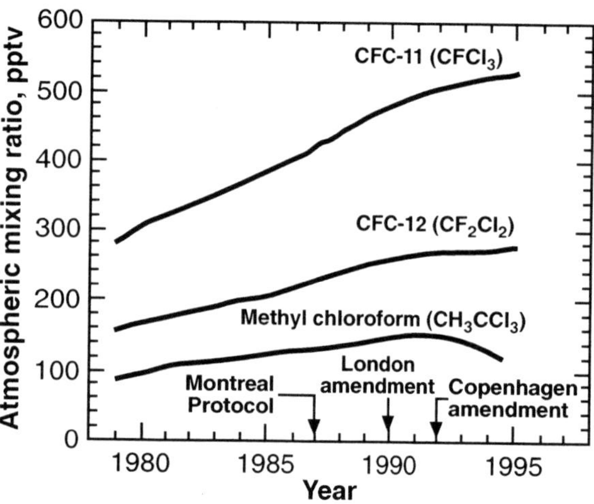

Figure 19. Tropospheric measurements of CFC-11 (CFCl₃), CFC-12 (CF₂Cl₂), and methyl chloroform (CH₃CCl₃) as a function of year. The atmospheric release rates of these compounds have been decreasing since 1991. The concentrations of CH₃CCl₃ are decreasing more rapidly than CFC-11 and CFC-12 because the atmospheric lifetime of CH₃CCl₃ is much shorter. From Ravishankara, A.R., and D.L. Albritton: Methyl chloroform and the atmosphere. Science 269, *1995, pp. 183–184. Reprinted by permission.*

levels should return to the pre–ozone hole levels and the ozone layer should recover.

Recent observations of CFCs in the atmosphere show that the rate of increase is now beginning to slow down (see Figure 19), indicating that the international agreements are working. In fact, one purely anthropogenic compound, methyl chloroform, is actually decreasing due to the controls. Methyl chloroform levels are dropping faster than those of CFC-11 and CFC-12 due to its shorter lifetime (5.4 years, compared with 50 years for CFC-11 and 102 years for CFC-12).

One interesting aspect of the above ozone protection process is that scientific advancements and policy decisions progressed in parallel and in coordination. As new scientific discoveries were made, scientists relayed them to the policy community, and policies evolved through the amendments to the Montreal Protocol. Such a process occurred because the original Vienna Convention specifically set forth a schedule and mechanism whereby the international scientific community would produce periodic "state of the ozone layer" assessments. Those reports directly fed into the policy negotiations and resulting decisions. The success of this international scientific/political feedback process is entirely unprecedented in the environmental arena.

Observations of a slowdown in growth rates of some CFCs are proof that the halogen challenge can be met by international agreements to control the compounds that contain chlorine and bromine. However, anthropogenic perturbations to the ozone layer can also be caused by other compounds. One possibility raised in the 1970s is related to the increased use of supersonic transport planes (SSTs) and their NO_x

Engineering Solutions to Ozone Depletion

Recently, some scientists have proposed various engineering solutions to the ozone-loss problem. One proposal has been to fill the ozone hole with ozone, like filling a pothole in the street. However, this is impractical on a mass basis (see questions). Another suggestion by Ralph Cicerone and colleagues was to inject propane, a hydrocarbon, into the stratosphere. The intended effect of adding propane was to scavenge the chlorine atoms. After inspecting

their calculations more closely, however, they determined that their proposal could make the ozone problem worse. So, in addition to the challenges of scaling up laboratory and theoretical solutions for the stratosphere, engineered solutions (sometimes called techno-fixes) to the ozone problem are risky unless all of the potential chemical ramifications are considered. Unforeseen reactions could also change the predicted outcome of techno-fixes.

emissions. Catalytic destruction of ozone due to the NO_x cycle, Reactions (5) and (6), was originally proposed by Paul Crutzen as a natural ozone-loss mechanism (because NO_x was produced naturally by oxidation of N_2O, a gas that is emitted by the Earth's biosphere). However, Harold Johnston pointed out that a proposed fleet of 500 SSTs would emit NO_x directly into the stratosphere, where the aircraft would fly to attain high speeds. This started a huge scientific and political debate on whether or not high-flying aircraft exhaust would damage the ozone layer. In 1975, the U.S. Congress decided not to build a fleet of SSTs, not because of their environmental impacts, but because they were not economically viable. More on the story of SSTs, CFCs, and ozone in the 1970s is given in *The Ozone Wars* by L. Dotto and H.I. Schiff. Johnston's impact on atmospheric chemistry and global public policy decisions earned him the 1997 National Medal of Science.

In recent years, newer versions of high-flying aircraft, now called high-speed civil transports (HSCTs), are being considered. As was the case with SSTs in the 1970s, the new HSCTs also would directly inject NO_x into the stratosphere, although the proposed new engine technology would result in much lower NO_x emissions. In addition, laboratory measurements of Reaction (18) on sulfuric acid surfaces indicate that the NO_x emissions would react on the background stratospheric sulfate aerosols to form nitric acid, essentially stopping ozone destruction by the NO_x cycle but increasing the importance of the ClO_x cycle. Another new twist in the HSCT debate is that the emissions from these aircraft could also increase the particle surface area available for reaction. Thus, detailed photochemical models including the heterogeneous reaction rates and particle formation calculations are needed to predict the effect of HSCTs on stratospheric ozone. In addition to the chemistry of HSCT emissions, the eventual distribution of the emissions in the stratosphere is important in assessing the impact of these aircraft on the ozone layer.

International research on HSCTs has raised the question of how the current subsonic fleet of aircraft is affecting the atmosphere. These planes usually fly in the troposphere except on polar routes, where the stratosphere is at a lower altitude. Currently, little is known about how the emissions from these planes change the chemistry of the upper troposphere and lower stratosphere and how widely the emissions are distributed (atmospheric dynamics). In addition to the chemical and dynamical questions, researchers are looking at how the emissions of subsonic aircraft affect cirrus (high-altitude) cloud formation, which is important for global radiation studies and climate (see module *Clouds and Climate Change*).

VI
Conclusion

Direct observations of the atmosphere have shown that the ozone hole occurs above Antarctica every spring and that chlorine from anthropogenic sources is responsible for this event. By studying the ozone hole phenomenon, scientists have determined that humans can cause environmental change on a global scale and have an impact on the Earth's future. In the case of the ozone layer, steps have been taken to avoid further environmental problems by regulating human activities. Because the effect of releasing CFCs is a global issue, international protocols have been established in a cooperative effort for the planet's future. The ozone hole event is observable every austral spring and will continue to form until the chlorine levels are reduced. The ozone hole has been considered a relatively simple problem where analysis of the underlying causes has resulted in the formation of a clear solution: restricted use of anthropogenic halogenated compounds. Unfortunately, another major global-scale environmental problem, climate change due to increasing greenhouse gases, is more complex and unlikely to be resolved in such a straightforward manner.

Future challenges in the study of the ozone layer lie in the area of global ozone loss over the midlatitudes, where most of the Earth's population lives. Currently, we do not understand the details well enough to provide a full explanation or to predict what may occur in the future. As with the increasing amount of CO_2 in the troposphere and how it relates to global warming, global ozone depletion needs long-term records and models that can represent all processes accurately. The linkage of ozone loss to tropospheric climate changes requires a better understanding of how the upper troposphere and lower stratosphere are coupled. For example, increasing concentrations of CO_2, CH_4, N_2O, and other greenhouse gases are believed to cause cooling of the stratosphere. Our present understanding of ozone-layer chemistry indicates that stratospheric cooling would cause more PSCs to form and the heterogeneous reactions on SSAs to occur more efficiently. Both of these factors would create more ozone loss. If, however, lower temperatures in the stratosphere change atmospheric circulation patterns, the effect of increasing greenhouse gases on the ozone layer would be harder to predict. Finally, we need to understand how emissions from fuel combustion affect the ozone layer so that we can make informed decisions about flying more aircraft in the stratosphere.

Problems and Discussion Questions

1. In terms of altitude, where are the troposphere and the stratosphere located? Where is ozone loss a problem? Where would increases in ozone be a problem?

2. How and where is ozone produced? Write out the relevant equations.

3. Write down a catalytic cycle for the destruction of ozone by chlorine.

4. How many ozone molecules can be destroyed by one chlorine atom?

5. Which of the following, if released in massive quantities on Earth, could lead to substantial ozone loss?
   HCl $\qquad$ $NaCl$ $\qquad$ CF_2Cl_2

6. Why are CFCs being regulated when NaCl is emitted from oceans in much greater abundance?

7. For what contributions did Molina, Rowland, and Crutzen win the 1995 Nobel Prize in chemistry?

8. Why did the U.S. ban on CFCs in aerosol spray repellents in the 1970s have relatively little impact on global levels of CFCs?

9. Since CFCs are produced in the Northern Hemisphere near the ground, why should we worry about their causing the Antarctic ozone hole?

10. What compounds are being used to replace CFCs, and why are they less harmful to the ozone layer?

11. What are halons, and why have they been banned?

12. What three environmental conditions are needed to cause massive ozone loss as observed in the Antarctic ozone hole?

13. Why is the ozone hole worse over the South Pole than over the North Pole?

14. What *natural* phenomenon causes ozone loss in the polar region to be worse some years than others? What *natural* phenomenon causes ozone loss in the global stratosphere to get worse?

15. Major explosive volcanic eruptions produce copious amounts of SO_2 and HCl. Explain why such eruptions cause a large increase in stratospheric sulfur but have little impact on stratospheric chlorine.

16. Why is ozone loss such a big problem for all life forms? What are the potential effects on humans of ozone loss?

17. Where are you more likely to get a sunburn, in the tropics or at midlatitudes? List three reasons for your answer.

18. Why do biologists looking for life on Mars want to look inside of rocks rather than on the surface of the planet?

19. How would each of the following emitted from aircraft potentially impact ozone?

    NO SO_2 H_2O

20. What government treaties are in place to help maintain the ozone layer? What impact did the discovery of the ozone hole have on the compounds regulated and the timeline?

21 The surface area of the Earth is 5×10^{18} cm^2. A typical ozone column is 300 DU.

   a. How many molecules of O_3 are there in the atmosphere?*

   b. What is the mass of O_3 molecules in the atmosphere? The weight of one O_3 molecule is 8×10^{-23} g.*

   c. There are 6 billion people on earth, with an average weight of 60 kg. What is the total mass of humans on earth?*

   d. Given the above masses, is it likely that you could replace the stratospheric O_3 layer by making O_3 at the surface of the Earth and transporting it up to the stratosphere?

Answers:

21a. 4×10^{37} 21b. 3×10^{15} g   21c. 3.6×10^{14} g

Glossary

Aerosol—a mixture of solid or liquid particles dispersed in a gas. Rain, snow, smoke, and hairspray are all examples of aerosols.

Anthropogenic—made by humans. Pizza is anthropogenic; broccoli can be considered either anthropogenic or natural; milk is natural; sea salt is natural.

Atmospheric dynamics—the air circulation pattern or how air moves in the atmosphere.

Austral—specific to the Southern Hemisphere, usually used when referring to seasons or stars.

Catalytic cycle—a set of reactions that can occur many times before stopping. At least one species is converted back to itself by other reactions in the cycle. Polar stratospheric ozone depletion occurs via a catalytic cycle, Reactions (7) and (9) through (11). In that catalytic cycle, a single chlorine atom can destroy 100,000 ozone molecules before another reaction can break the chain of reactions.

Chlorine activation—a process where inert chlorine reservoir molecules, $ClONO_2$ and HCl, are converted into active chlorine radicals, Cl and ClO, that can destroy ozone in a catalytic cycle.

Chlorofluorocarbons (CFCs)—anthropogenic, carbon-based compounds containing chlorine and fluorine that are used as refrigerants, blowing agents, and cleaning solvents. Their production is now being limited because the chlorine from these molecules eventually destroys stratospheric ozone.

Column amount—the total number of ozone molecules above a square centimeter of the Earth's surface.

Denitrification—a process where nitrogen species are permanently removed from the stratosphere. Denitrification is observed when the measured amount of nitrogen species is less than predicted.

Halons—anthropogenic, carbon-based compounds containing bromine that are used in fire extinguishers. Like CFCs, their production is currently being regulated because the bromine in these compounds eventually destroys stratospheric ozone.

Heterogeneous—refers to two physical states or phases, like gas and solid or gas and liquid. The opposite of heterogeneous is homogeneous, where the physical state is the same. For example, ice in water is a heterogeneous (solid and liquid) system. Cranberry-apple juice is an example of a homogenous (one-phase) system. Sodas can be either: they are heterogeneous systems when fresh (liquid with gas bubbles) and homogeneous when flat.

Hydrochlorofluorocarbons (HCFCs)—anthropogenic, carbon-based compounds containing chlorine, fluorine, and hydrogen that are being developed to temporarily replace CFCs. They react in the troposphere, so they are "safer" than CFCs for the ozone layer. As with CFCs their production will eventually be stopped, and other compounds are being developed to replace them.

Inert—nonreactive. Couch potatoes are inert.

Latitude—an indicator of a distance from the equator. The equator is at 0° (° = degrees) latitude and the poles are at 90° latitude (90°N is the North Pole and 90°S is the South Pole). Low latitudes are near the equator and the tropics, midlatitudes are where most of the world's population lives, and high latitudes are near the poles.

Photochemical reactions (or photolysis)—reactions that are light-initiated. Photosynthesis is a photochemical reaction where carbon dioxide reacts with water in the presence of light to produce oxygen and sugars, the material of new plant growth.

Polar stratospheric clouds (PSCs)—clouds formed in the stratosphere during the cold polar winter. Heterogeneous reactions of chlorine compounds on the surfaces of these clouds play an important role in ozone destruction.

Stratosphere—the part of the upper atmosphere where the ozone layer exists. In the stratosphere, the temperature increases with increasing altitude.

Stratospheric sulfate aerosols (SSAs)—small, sulfuric acid particles that exist in the stratosphere globally. Heterogeneous reactions of nitrogen and chlorine compounds on SSAs could be fostering global ozone loss.

Tropopause—the altitude where the troposphere and stratosphere meet. The tropopause height varies with season and latitude.

Troposphere—the part of the atmosphere closest to Earth. We breathe the air in the troposphere, and weather is controlled by tropospheric air motions.

Ultraviolet (UV) radiation—high energy light with wavelengths between 200 and 400 nm. Table 1 contains details on regions of the UV spectrum.

Suggested Additional Reading

Arctic ozone: AASE II observations. *Science* 261 (special issue), 27 August 1993.

The changing atmosphere: Implications for mankind. *Chemical and Engineering News* 64(47) (special issue), 24 November 1986.

Dotto, L., and H.I. Schiff: *The Ozone Wars*. Doubleday, Garden City, New York, 1978.

National Research Council: *Atmospheric Effects of Stratospheric Aircraft: An Evaluation of NASA's Interim Assessment*. National Academy Press, Washington, D.C., 1994.

Toon, O.B., and R.P. Turco: Polar stratospheric clouds and ozone depletion. *Scientific American*, June 1991, 68–74.

Turco, R.P. : The stratospheric ozone layer. In *Earth Under Siege: Air Pollution and Global Change*, Oxford University Press, New York, 1996.

Wayne, R.P.: Ozone in the earth's stratosphere. In *Chemistry of Atmospheres* (2nd ed.), Oxford University Press, New York, 1991.

World Meteorological Organization (WMO) Global Ozone Research and Monitoring Project: *Atmospheric Ozone: 1985*. No. 16, WMO, Geneva, Switzerland, 1985.

WMO Global Ozone Research and Monitoring Project: *Scientific Assessment of Ozone Depletion: 1989* (No. 20), *1991* (No. 25), *1994* (No. 37), and *1998* (No. 44), WMO, Geneva, Switzerland.

Other modules in the UCAR Global Change Instruction Program Available from University Science Books, 55D Gate Five Road, Sausalito, CA 94965

Ennis, C., and N. Marcus: *Biological Consequences of Global Climate Change*

Rosencranz, A.: *International Environmental Law and Policy*

Streete, J.: *The Sun-Earth System*

Websites:

www.unep.ch/ozone/

www.unepie.org/ozonaction.html

Index